Sergey Brin

Sergey Brin

Groundbreaking Google Founder

MATT DOEDEN

LERNER PUBLICATIONS ◆ MINNEAPOLIS

Lerner Publications Company
An imprint of Lerner Publishing Group, Inc.
241 First Avenue North
Minneapolis, MN 55401 USA

For reading levels and more information, look up this title at www.lernerbooks.com.

Image credits: Steve Jennings/Getty Images for MerchantCantos, p. 2; Taylor Hill//Getty Images, p. 6; Kim Kulish/Corbis/Getty Images, p. 8; INTERFOTO/Alamy Stock Photo, p. 10; Dave Buresh/ The Denver Post/Getty Images, p. 12; Barry John Stevens/Fairfax Media/Getty Images, p. 13; jejim/Shutterstock.com, p. 15; MediaNews Group/Bay Area News/Getty Images, p. 17; Jeramey Lende/Shutterstock.com, p. 19; JOKER/Martin Magunia/ullstein bild/Getty Images, p. 20; Tim Mosenfelder/Getty Images, p. 22; Asahi Shimbun/Getty Images, p. 23; Kimberly White/Corbis/ Getty Images, p. 23; im Kulish/Corbis/Getty Images, p. 25; Smith Collection/Gado/Getty Images, p. 26; Spencer Platt/Getty Images, p. 28; AP Photo/Ben Margot, p. 29; Justin Sullivan/Getty Images, p. 33; FloridaStock/Shutterstock.com, p. 34; Sundry Photography/Shutterstock.com, p. 36; Ian Tuttle/Breakthrough Prize/Getty Images, p. 37; AP Photo/Paul Sakuma, p. 38.

Cover: Kelly Sullivan/Getty Images.

Main body text set in Rotis Serif.
Typeface provided by Adobe.

Designer: Viet Chu
Lerner team: Sue Marquis

Library of Congress Cataloging-in-Publication Data

Names: Doeden, Matt, author.
Title: Sergey Brin : groundbreaking Google founder / Matt Doeden.
Description: Minneapolis : Lerner Publications, [2022] | Series: Gateway biographies | Includes bibliographical references and index. | Audience: Ages 9–14 | Audience: Grades 4–6 | Summary: "Sergey Brin was born in Moscow, Russia, and moved to the United States in 1979 to avoid Jewish persecution. With Larry Page, Brin co-founded Google and became one of the world's wealthiest tech entrepreneurs"– Provided by publisher.
Identifiers: LCCN 2020000496 (print) | LCCN 2020000497 (ebook) | ISBN 9781541596740 (hardcover) | ISBN 9781728413549 (paperback) | ISBN 9781728400334 (ebook)
Subjects: LCSH: Brin, Sergey, 1973- –Juvenile literature. | Google (Firm)–Juvenile literature. | Internet industry–United States–Juvenile literature. | Webmasters–United States–Biography– Juvenile literature. | Businesspeople–United States–Biography–Juvenile literature.
Classification: LCC QA76.2.B75 D64 2020 (print) | LCC QA76.2.B75 (ebook) | DDC 338.7/6102504092 [B]–dc23

LC record available at https://lccn.loc.gov/2020000496
LC ebook record available at https://lccn.loc.gov/2020000497

Manufactured in the United States of America
1-47782-48219-4/12/2021

CONTENTS

Sergey Brin's Google is one of the best-known technological inventions of our time.

A crowd of more than five thousand people filled the Moscone Center in San Francisco, California, on June 27, 2012. They were there for Google's annual conference for web developers. They eagerly awaited the event's opening presentation.

Sergey Brin, the cofounder of Google, stepped onto the stage wearing an unusual pair of glasses. In some ways, they looked like ordinary reading glasses, but a small electronic device was affixed to one of the lenses. Brin was wearing one of the first pairs of Google Glass, an exciting and innovative product that he and his team at Google had created.

The glasses included a built-in computer display, cameras, and an internet connection. Users could record everything they saw and heard and even livestream the scene directly to the internet. Google, which had started as an internet search engine, was growing and expanding into a wide range of new high-tech products. Google Glass was one of them.

"We've got something pretty special for you," Brin told his audience. As he spoke, a screen behind him showed an airship hovering over the streets of San Francisco. A team of skydivers, each wearing Google Glass, was ready to jump and stream the experience in real time. "This could go wrong in about 500 different ways," Brin cautioned. A rumble of laughter swept through the auditorium.

Brin and the crowd watched excitedly as the view shifted to the Google Glass worn by one of the jumpers. The city of San Francisco stretched out below. A roar filled the auditorium as the team jumped. As Brin—a recreational skydiver himself—narrated, the jumpers glided in high-tech wingsuits, headed straight for the roof of the Moscone Center.

Sergey Brin watches a live feed as skydivers wearing Google Glass jump from a plane over San Francisco.

The jumpers opened their parachutes, landed safely on the roof, and within minutes joined Brin on the stage. This was the kind of spectacular, headline-grabbing demonstrations that Brin loved, and it had gone off perfectly. Although Google Glass itself did not enjoy great success, its introduction helped mark a period of transition in Google, one of the world's biggest companies. Google had begun life as a web search business. But Brin and his fellow cofounder, Larry Page, were no longer content just to produce web-based products. They were branching out into new forms of technology and revolutionizing what was possible.

FROM MOSCOW TO MARYLAND

Sergey Mikhaylovich Brin was born on August 21, 1973, in Moscow, Russia. At the time, Russia was part of the Soviet Union. This powerful nation denied its citizens many rights and freedoms. Sergey's parents, Mikhail and Eugenia Brin, had both graduated from the University of Moscow. Sergey's father was an economist for a government agency. His mother worked in a research lab. The young family appeared to be well set for their future. But there was one catch. The Brins were Jewish. At the time, anti-Semitism—prejudice against Jewish people—was high in the Soviet Union. Because the Soviet government discriminated against Jews, Mikhail Brin worried that his son's opportunities in school and future employment would be limited.

Sergey Brin was born in Moscow, Russia, part of the former Soviet Union.

In 1977 Mikhail Brin attended a conference in Warsaw, Poland. While there, he met people who lived outside the Soviet Union. He learned that people in many other countries didn't suffer under controlling and oppressive governments. They had more choices and opportunities in life. Mikhail Brin made a decision that he hoped would give his son a better life. He would apply for an exit visa. This official document would allow the Brins to leave the Soviet Union and become citizens of another country.

It wouldn't be easy, though. The Soviet Union did not always approve applications for exit visas. People who applied and were denied were called refuseniks. The Soviet government made their lives very hard. It often had them fired from their jobs and denied them new ones. Even applying to leave the country came at a great risk of punishment. At first, Sergey's mother wasn't sure that applying was worth it. But in the end, she agreed. The Brins applied for their exit visas in September 1978.

Their application had immediate consequences. His parents lost their jobs. His father made money by translating books from Russian into English. It was slow and tedious work, but it kept the family afloat.

The waiting period was tense. No one knew whether their application would be accepted. They didn't know how long it might take. All they could do was survive. Many Soviet Jews were requesting exit visas. For many of them, the wait dragged on for years. But the Brins were lucky. In May 1979, their application was approved. They were allowed to leave the country.

This marked the beginning of a long journey for six-year-old Sergey and his parents. "We were in different places from day to day," Sergey recalled. The family went from Vienna, Austria, to Paris, France, as they searched for a place to call home. Finally, in October, they boarded an airplane headed for the United States. They landed in New York City and settled in a small, rented home in Maryland. Mikhail Brin took a job teaching mathematics at the University of Maryland. The Jewish community in Maryland helped support the family financially while they struggled to get on their feet.

Sergey's parents enrolled him at Paint Branch

Montessori School, a private school in Adelphi, Maryland. He also studied Hebrew at Mishkan Torah Synagogue, a Jewish congregation in Greenbelt, Maryland. School was difficult for the six-year-old. He spoke only a little English, and he had a heavy Russian accent. "It was a difficult year for him, the first year," his mother recalled. "We . . . had been told that children are like sponges [quickly absorbing a new culture], that they immediately grasp the [new] language and have no problem, and that wasn't the case."

Slowly, he began to fit in. He was a curious and

People began using personal computers on a wide scale in the 1980s. These boys won a computer contest at their school.

focused learner, although not an outstanding student. Like his father, he had a knack for mathematics. Often he found his schoolwork too easy—especially in math—so he tackled more advanced concepts at home with his parents. When he was nine, Sergey received a personal computer from his father. Sergey was fascinated by it. At the time, home computers were still a relatively new technology. Sergey surprised his teacher by bringing in a school assignment that he'd typed on his computer and printed out.

The first desktop computers were large, bulky, and slow.

Still nothing about young Sergey hinted at what he would become. "Sergey wasn't a particularly outgoing child," said Patty Barshay, the director of Sergey's school. "But he always had the self-confidence to pursue what he had his mind set on. . . . He was interested in everything [but] I never thought he was any brighter than anyone else."

The Brin family expanded in the United States. In 1987 Eugenia Brin gave birth to another boy, Sam. Fourteen-year-old Sergey then had a little brother. When Sergey was sixteen, he joined his parents and other advanced mathematics students on a trip to Moscow. It was the first time he'd been back to his birth nation since the family had fled. Seeing the Soviet Union, where people struggled with poverty under strict government control, made a big impact on young Sergey. He even threw pebbles at a police car in anger. The police officers were angry, but his parents convinced them to let Sergey go with a warning.

The trip gave Sergey a new appreciation for the opportunities he had in the United States, where citizens are guaranteed many rights and freedoms, such as freedom of speech. On the second day of the trip, Sergey pulled his father aside. "Thank you for taking us all out of Russia," he said.

FINDING A FOCUS

In his teen years, Sergey attended Eleanor Roosevelt High School. There, he grew more and more driven to get a good education. Early on, he planned to follow in his father's footsteps. He wanted to be a mathematician.

At Roosevelt High, Sergey had loaded up on classes that offered college credit. In 1990 he enrolled at the University of Maryland. Because he had taken so many college-level classes, he already had nearly a year of college credit. That allowed him to earn his college degree in three years rather than the usual four.

Brin followed his passions at college. Over three years, he studied mathematics and computer science. He graduated with high honors and earned a National

Science Foundation scholarship for graduate school.

Brin had no doubt where he wanted to go next: Stanford University in Stanford, California, then at the heart of the growing computer technology industry. In 1993 Brin enrolled at Stanford. He planned to pursue a master's degree and a PhD in computer science. He didn't yet know exactly what he would be doing with his life, but he believed that his future was somewhere in the computer industry. So he packed his bags and headed to California.

Brin was also gaining real-world experience. In the summer of 1993, he interned at Wolfram Research, a software company. While there, he worked as part of the team that developed a computing system called

In the 1990s, students at Stanford University made groundbreaking advances in computer technology.

Mathematica. It was Brin's first real taste of what working at a large-scale technology company was like.

Brin attended Stanford when businesses and individuals were just starting to use the internet. Brin became fascinated with the amount of information on the web and on how that information could be gathered and organized. He started writing computer programs that automatically collected images from popular websites and stored the information on his computer.

SEARCH ENGINES

IN THE EARLY 1990S, THE WORLD WIDE WEB WAS DISORGANIZED AND HARD TO USE. Finding information was difficult. To help other users find websites, researchers spent hours looking through hundreds or even thousands of sites. Then they compiled indexes, or lists of sites. They made indexes of sites about basketball and indexes of sites about US politics, for example.

In 1993 programmers created the first web search engine, Wandex. Other search engines quickly followed. They included Lycos and Infoseek in 1994 and Yahoo! and AltaVista in 1995. These search engines used software called crawlers. They indexed the web by searching for key words—critical words that appeared on a web page. It took seconds instead of hours. Early search engines made it easier to find information, but they did not always lead users to the information they were seeking.

Tech companies such as Apple built large office complexes in Silicon Valley.

A SHARED VISION

In March 1995, Brin met Larry Page at a meeting of computer science graduate students at Stanford. The two bright students didn't click at first. "When we first met each other, we thought the other was really obnoxious," Page recalled. "Then we hit it off and became really good friends."

Brin and Page discovered that they shared a common interest in the science of processing information. At the time, internet search engines were often ineffective. They worked by searching for key words. But the results they returned to users were often frustrating. The websites they uncovered didn't always provide the type of information users had set out to find. Page was working on a project he thought could change that. Instead of key words, he wanted to focus on how pages linked to one another. He reasoned that if a lot of other websites linked to a particular page, it probably contained useful information. And if few or no sites linked to the page, it was probably not very useful.

Brin was intrigued by Page's idea and joined his new friend's project. "We started working really, really hard on it," Page said. "We worked through holidays, and worked many, many hours a day."

The pair cowrote a series of academic papers about data mining, or sifting through large amounts of data to get smaller sets of the most useful data. In 1996 they set their ideas into motion, creating their own internet search engine, which they called BackRub. The name was a play on the term *backlinks*, or links from one web page to another.

BackRub used backlinks as the foundation for ranking search results. The sites that showed up at the top of search results were those with the most backlinks, which meant that other website operators found the sites to be useful. The system was a revolution in search engine technology. But as students, Brin and Page lacked the money to start their own search engine company. They ran their search engine out of Stanford's Computer Science Department. The search engine was available only to Stanford students.

Thanks to Brin and Page's research and development work, the websites you see at the top of your search results are usually the most relevant sites.

In the fall of 1997, Brin and Page were ready to bring their idea to the broader web. So the pair sat down to think of a new name for their search engine. BackRub didn't seem right. It didn't sound like a groundbreaking technology business. Fellow student Sean Anderson suggested that they use the name Googolplex. This word comes from a number called a googol. It is written with the figure 1 followed by one hundred zeroes. A googolplex is much larger: it is written as the figure 1 followed by a googol of zeroes. Brin liked Anderson's idea but thought they should use the shorter name Googol instead of Googolplex.

Sergey Brin (*left*) and Larry Page hold a poster showing the Google logo.

Anderson checked whether the domain name Googol had already been taken by another website. But when he searched for it, he misspelled the word. Instead of *googol*, he searched for *google*. Not surprisingly, the misspelled name was not being used by another website. The group quickly realized the error. But Brin and Page thought that Google was a good name too. So they registered the name Google.com with an international listing of domain names. A new company was born. Page's Stanford dorm room became Google's computer center, while Brin's room was its business office.

BUILDING GOOGLE

Brin's life revolved around building the Google search engine. At first, he and Page thought of it as a project that would help them earn their PhDs. But they began to see that it could become something much bigger. The pair thought they could sell their idea to an existing search engine. They pitched their program to AltaVista, Yahoo!, and AOL. Their asking price was $1 million. But no one was interested.

Brin and Page couldn't find a buyer for their idea. But they did find an investor. Andy Bechtolsheim, cofounder of the computer company Sun Microsystems, saw promise in what they'd built. In 1998 Bechtolsheim wrote a check for $100,000 to Google, Inc. That investment made Bechtolsheim a part owner of the business and also provided the money needed to start up operations. There was one problem. Brin and Page had not yet turned Google into a legal corporation. So the check sat uncashed for several weeks while they scrambled to file the proper government paperwork.

Flush with new funding, they got to work. Google, Inc., moved its computer servers out of Page's dorm room and into rented space in the garage of friend Susan Wojcicki. It was a haphazard office, with desks made out of sawhorses and old pine doors, as well as a Ping-Pong table. But it was a step up from the dorms. Page was the company's chief executive officer while Brin was the president.

Brin married 23&Me founder Anne Wojcicki in 2007.

That year Brin completed his master's degree in computer science. But he quit the PhD program so that he could give Google his full attention. This was not welcome news to Brin's parents, who didn't quite understand what he was doing with Google. But Brin had a vision, and he couldn't be talked out of it.

Brin may have been focused on Google, but someone captured his attention around this time: Susan Wojcicki's sister, Anne Wojcicki. The pair hit it off and started dating. They had a lot in common. Both were interested in technology. In her job as a health-care business analyst, Anne Wojcicki focused on biotechnology (the use of technology in medicine). The couple discovered a shared interest in yoga as well as outdoor activities. Brin loved to swim and dive, while Wojcicki was an avid biker.

"Sergey's amazingly creative," Wojcicki said years later. "I mean, that's the fun of him. The stuff that comes out of his mouth sometimes is just extraordinary. He really genuinely sees the world in a different way. He's also not worried about what people think of him."

RIDING THE BOOM AND WEATHERING THE BUST

Brin and Page started Google in the middle of the dot.com boom. From about 1994 to 2000, web-based companies were exploding. The internet was still young, but users were excited about its potential to change the way people shopped, communicated with one another, searched for information, and operated businesses.

In 1998 Google was still a beta project—users could do searches on the site, but it wasn't fully finished. Yet computer experts said that Google's unique system of ranking search results made it better than established search engines. As more users discovered its effectiveness, Google.com grew quickly. Within a few months of launching, the new search engine was performing ten thousand searches a day. The media began to notice. Publications such as *USA Today* wrote articles about the website. One of Google's biggest breaks came in December, when *PC Magazine* included Google in its lists of the top 100 websites for 1998.

Technology businesses boomed at the turn of the twenty-first century.

Also in December 1998, an article in the online magazine Salon praised the new search engine. "There is a better way to build a search engine. And a Silicon Valley start-up company with the unlikely name of Google.com is showing the way," wrote Salon's Scott Rosenberg. "Since discovering Google a few weeks ago, I've been so impressed with its usefulness and accuracy that I've made it my first search stop."

Publicity like that fueled Google's rapid growth. More and more people were using it for web searches. The company began hiring employees to help build and operate the search engine. By March 1999, the company had outgrown the garage. So Brin and Page moved into new offices in Palo Alto, California. By the summer of 2000, the site was performing about eighteen million searches per day. The cost of running Google's servers grew and grew. The company operated with money from investors such as Andy Bechtolsheim. But it also needed to earn money to stay in business.

Since the earliest days of the internet, search engines have always been free to use. When Google was starting out, most major search engines (and many other websites) made money by posting large, colorful advertisements on their pages. Advertisers paid hosting sites a small fee every time a user clicked on an ad, which took the user to the advertiser's website.

Brin hated ads on web pages. He wanted Google to have a clean look. He wanted the site to be about

Page (*left*) and Brin show off the servers at Google's Mountain View headquarters.

searching, not selling. But he also knew that his company needed to start generating money. So he tried something new. He added "sponsored search results" to the site. Instead of running ads on Google, advertisers paid to have their links appear at the top of lists of search results, in an area clearly marked as sponsored (paid-for) links. For example, a search for "travel to Hawaii" might bring up sponsored links to the websites of airlines and hotel chains. Sponsored search results were an elegant solution to Google's money problems. The advertisers who paid for the links gave Google the steady stream of income it desperately needed while allowing Brin's clean, uncluttered vision of the site to continue.

Brin's stroke of advertising genius came just in time. Starting in spring of 2000, the dot.com boom began to fade. Many websites went out of business—including some of Google's competitors. But even as many sites struggled, Google was thriving. Slowly, the site was becoming the world's go-to search engine. When people needed information about anything, all they had to do was go online and google it.

Google headquarters is called the Googleplex.

JUST GOOGLE IT

IN THE LATE 1990s, GOOGLE WAS SIMPLY A COMPANY NAME.
But soon Google employees started using *google* as a verb. When someone wanted to find information online, they "googled it." The phrase began to spread through popular culture. For instance, in 2002 a character on the TV show *Buffy the Vampire Slayer* asked another character, "Have you googled her yet?" Within a few years, "google it" saw such widespread use that it all but replaced any other way of describing a web search. The *Oxford English Dictionary* officially added the verb *google* in 2006.

GOING PUBLIC AND STAYING PRIVATE

As Google grew, honors rained down on Brin and Page. In 2002 the *MIT* (Massachusetts Institute of Technology) *Technology Review* named them to its list of the top 100 innovators under the age of thirty-five. Two years later, they were awarded the Marconi Prize, one of the biggest awards in engineering.

The duo continued to expand and innovate. They branched out into new markets. The company opened its first international office, in Japan, in 2001. By 2002 Google operated in seventy-two languages. Meanwhile, Google developed new types of web services. For example, it created Google Maps, which offers detailed online street maps. Using images taken from satellites, the company created Google Earth, which gives users a close-up, bird's-eye view of just about any place on the planet.

In a few short years, Google had become one of the biggest businesses online. Brin and Page wanted to run an employee-friendly company. In 2004 they opened a sprawling headquarters in Mountain View, California. Playing on the original name for the business, they called it the Googleplex. They believed that happy, relaxed employees were a key to a company's success. The Googleplex provided employee perks such as pool tables, free food, free laundry services, and even free massages. Employees could dress as they pleased and bring pets to work.

Google employees have created products such as Gmail, Google Earth, and Google Maps.

The Google logo is familiar to computer users around the world.

Brin said that learning to lead the growing company was one of his biggest challenges. "I think managing people, and being emotionally sensitive, and all the skills you learn in terms of communication and keeping people motivated, that has been a challenge," he noted. "I have enjoyed learning that, but that's important, and a hard thing to learn."

All the hard work built up to one of the biggest days in the history of Google. Brin and Page had already sold shares in their business to private investors. But they were ready to sell shares on the stock market. That way, anyone who wanted to own a portion of the business could buy shares in Google. This process, called going public, started with an initial public offering, or IPO.

Google's IPO, on August 4, 2004, made waves in the stock market. Brin and Page sold millions of shares of their company. Over the following months, more people wanted to invest in the business. The price of Google stock rose higher and higher. Because Brin owned a large portion of the stock himself, his wealth grew along with the company. By October 2004 he was worth $6.5 billion. He was one of the wealthiest people in the world.

Brin was becoming a public figure. But he valued privacy in his personal life. One area of his life he kept especially private was his relationship with Anne Wojcicki. In May 2007, the couple was ready to get married. But they didn't want uninvited reporters and photographers at the wedding. So they planned it in secret. Even the wedding guests didn't know where it would be held. The couple and all their guests boarded a jet that flew them to a private island in the Bahamas. Brin and Wojcicki, along with many of their guests, put on swimsuits and swam out to a sandbar, where the couple exchanged vows.

The couple quickly started building a family. They welcomed a son, Benji, in 2008. Three years later, their daughter, Chloe, was born. Instead of giving the children either of their last names, they combined the names to make the last name Wojin.

Both Brin and Wojcicki worked hard. Wojcicki cofounded and worked as chief executive officer of 23andMe, a company that analyzes genes (chemical markers, inherited from the birth parents, that determine a person's traits). But the couple agreed that putting energy into their children was important. Both made an effort to spend time as a family, especially in the evenings.

A NEWFOUND RISK

ANNE WOJCICKI'S COMPANY, 23ANDME, SELLS TESTS THAT ANALYZE PEOPLE'S GENES. Genes are passed down from birth parents to children, one generation to the next, over thousands of years. By testing genes, 23andMe helps people find out what part of the world their ancestors came from.

The company's tests also look for genes that can put someone at risk for getting certain diseases. When 23andMe tested Sergey Brin's genes, it found that he is at a high risk for developing Parkinson's disease. Brin's mother has Parkinson's, and she likely passed on the gene to him. This disease affects the nervous system. People with Parkinson's disease gradually lose the ability to control their muscles. Movement becomes very difficult.

Having the Parkinson's gene doesn't mean that Brin has the disease. It just means he's at a higher risk of getting it than the average person. But he made the results of his test public to raise awareness about the disease. The more people know about Parkinson's disease, the more they will support research to find new medicines and treatments for Parkinson's patients.

Sergey Brin (*left*) promotes clean energy alternatives. Here, he and Larry Page charge an electric car.

GROWTH AND CHANGE

For years Brin had focused on creating and building his search engine. As Google continued to grow into the world's most used website, Brin's life changed. Google still was a huge part of it. Brin helped the company create new products such as Gmail and the Chrome web browser. He also helped lead a new department, Google X, which focused on high-tech, futuristic projects such as Google Glass. But he also spent time on other interests.

Brin was concerned about climate change. The burning of fossil fuels (coal, oil, and natural gas) adds extra greenhouse gases, especially carbon dioxide, to the atmosphere. Greenhouse gases trap the sun's heat, causing global temperatures to rise. This warming has led to fierce storms, droughts, wildfires, and other natural disasters. Scientists predict more such disasters as global temperatures continue to climb.

To help fight climate change, Brin has invested in companies working on clean energy solutions. For example, he has invested in solar energy technology and wind farms, which generate power without releasing carbon dioxide into the air. Brin and Page started a yearly conference, nicknamed Google Camp, where many

Google employees take a volleyball break at company headquarters.

prominent people, including business leaders, politicians, and Hollywood celebrities, gather to discuss pressing world problems. One year the camp focused on the dangers of—and solutions to—climate change. Brin also became more active in charities, including those that supported research into Parkinson's disease.

Some animals are losing their habitats due to climate change.

DON'T BE EVIL

AROUND 2000 BRIN AND PAGE ADOPTED A COMPANY MOTTO FOR GOOGLE: "DON'T BE EVIL." They believed that many companies, including big web search engines, were in business just to make money rather than to help people. The motto served as a reminder to them that doing the right thing was better than making short-term profits.

Google has struggled with its "Don't be evil" rule over the years. For instance, starting in 2005, Google made a lot of money by offering its search engine in China. But like the former Soviet Union, the Chinese government is repressive. It restricts internet access to control what Chinese people can search for, write about, and read online. When people in China made a Google search, they didn't get the full search results. For instance, if an article was critical of the Chinese government, Chinese censors blocked it. Brin's childhood in the Soviet Union had given him a taste of life under that type of government. He and others at Google felt that despite all the money they were making in China, they didn't want to work with a country that censored free speech. So Google decided to stop doing business in China in 2010.

Brin also turned some of his attention to political causes. In 2017 he was among a group of protesters who gathered at San Francisco International Airport. They protested the immigration policies of US president Donald

Some protesters have spoken out in favor of allowing those from oppressive governments to come to the United States.

Trump. Trump wanted to greatly reduce the number of immigrants allowed into the United States. He didn't want the nation to welcome people fleeing from oppressive governments.

Brin understood the plight of these immigrants. "I'm here because I'm a refugee [someone who flees to a new country to escape danger or persecution]," he explained on his Twitter account.

Later, Brin spoke more about why immigration was such an important issue to him. "This country was brave and welcoming and I wouldn't be where I am today or have any kind of the life that I have today if this was not a brave country that really stood out and spoke for liberty," he said. "I came here to the US at age six with my family from the Soviet Union. . . . The US had the courage to take me and my family in as refugees."

Change came to Brin's personal life as well. Even though he was married, in 2014, he began a romantic relationship with Amanda Rosenberg, a Google employee who was working on Google Glass. Their affair became headline news. He and Wojcicki quickly divorced, although they remained on friendly terms and continue to parent their two children together.

The romance between Brin and Rosenberg quickly faded. But Brin was not alone for long. Around 2016 he began dating businesswoman Nicole Shanahan. The pair soon married. Shanahan gave birth to the couple's first child, a daughter, in 2018.

Nicole Shanahan and Sergey Brin attend an awards ceremony in 2019.

Brin (*left*) and Page no longer run daily operations at Google.

STEPPING AWAY

In 2015 Brin and Page announced that Google and its many divisions would be part of a new parent company, called Alphabet. Brin served as the president of the newly formed company.

In December 2019, it was time for Brin and Page to leave Alphabet altogether. They announced together that they were stepping down from their roles. They would remain involved in the company as its two largest shareholders, but their time running the company's day-to-day operations was over. They thought that other leaders were better suited to manage the business. "We've never been ones to hold on to management roles when we think there's a better way to run the company," they explained.

By 2020 Sergey Brin was richer than ever, with a net worth of more than $60 billion. He was one of the wealthiest immigrants in the United States. In 2021 Brin's net worth increased to over $89 billion. *Forbes* listed him as the ninth richest person in the world.

Brin's time running Google is over, but he continues to invest in technology that could change the world. This work includes a project to build high-tech airships. Brin envisions the ships being used to deliver food and supplies to people in remote areas, where it's difficult for airplanes to land. Having revolutionized the web, Sergey Brin continues to push the boundaries of technology.

IMPORTANT DATES

1973   Sergey Mikhaylovich Brin is born on August 21 in Moscow, Russia.

1978   The Brin family applies for exit visas to leave the Soviet Union.

1979   Brin leaves the Soviet Union with his family. They settle in Maryland.

1989   Brin briefly returns to Moscow on a trip with his parents.

1990   Brin graduates from high school and enrolls at the University of Maryland.

1993   Brin enrolls at Stanford University with plans to earn a PhD in computer science. The first search engine, Wandex, is created.

1995   Brin meets Larry Page. The two share an interest in collecting and processing information on the World Wide Web.

1996   Brin and Page complete their first search engine, called BackRub, which is available to fellow Stanford students.

1997 Brin and Page rename their search engine Google.

1998 Andy Bechtolsheim invests $100,000 in Google. Brin and Page officially launch their company and move operations into a friend's garage. *PC Magazine* lists Google as one of its top 100 websites of the year.

2000 Google allows advertisers to buy sponsored links, providing the company with an income and allowing it to grow.

2004 Brin and Page open a massive new headquarters, called the Googleplex. Google sells shares on the stock market, making billions of dollars for Brin, Page, and their early investors.

2017 Brin is among a group that protests the immigration policies of US president Donald Trump.

2019 Brin and Page step down from their positions at Google.

2021 *Forbes* lists Brin as the ninth wealthiest person in the world.

SOURCE NOTES

7–8 S. Pangambam, "Google I/O 2012 Keynote—Day 1 (Full Transcript)," Singju Post, July 12, 2014, https://singjupost .com/google-io-2012-keynote-day-1-full-transcript /?singlepage=1.

11 Mark Malseed, "The Story of Sergey Brin,"*Moment,* February 2007, https://web.archive.org /web/20130121055147/http://www.oldsite.momentmag. net/moment/issues/2007/02/200702-BrinFeature.html.

12 Malseed.

13 Malseed.

14 Malseed.

17 David A. Vise and Mark Malseed, *The Google Story* (New York: Bantam-Dell, 2005), 16.

18 Vice and Malseed.

22 Zoë Bernard, "23andMe CEO Anne Wojcicki's Defense of Her Ex-Husband's Weird FiveFinger Shoes Is a Glimpse into How She Thinks," Business Insider, November 20, 2017, https://www.businessinsider.com/23andme-founder -anne-wojcicki-speaks-about-sergey-brin-2017-11.

24 Scott Rosenberg, "Let's Get This Straight: Yes, There Is a Better Search Engine," Salon, December 22, 1998, https:// www.salon.com/1998/12/21/straight_44/.

27 Robinson Meyer, "The First Use of 'to Google' on

Television? *Buffy the Vampire Slayer*," *Atlantic*, June 27, 2014, https://www.theatlantic.com/technology/archive /2014/06/the-first-use-of-the-verb-to-google-on -television-buffy-the-vampire-slayer/373599/.

29   "Sergey Brin," American Academy of Achievement, accessed February 7, 2020, https://www.achievement .org/achiever/sergey-brin/.

35   Liam Tung, "Google Erases 'Don't Be Evil' from Code of Conduct after 18 Years," ZDNet, May 21, 2018, https:// www.zdnet.com/article/google-erases-dont-be-evil-from -code-of-conduct-after-18-years/.

36   T. C. Sottek, "Google Co-founder Sergey Brin Joins Protest against Immigration Order at San Francisco Airport," Verge, January 28, 2017, https://www.theverge .com/2017/1/28/14428262/google-sergey-brin.

36–37 Kate Samuelson, "Google's Co-founder Criticizes President Trump's Refugee Ban in Passionate Speech," *Fortune*, January 31, 2017, https://fortune.com /2017/01/31/sergey-brin-donald-trump/.

38–39 Kaya Yurieff, "Google Co-founders Larry Page and Sergey Brin Stepping Down as Alphabet Executives," CNN, December 4, 2019, https://www.cnn.com/2019/12 /03/tech/alphabet-google-co-founder-larry-page-step -down/index.html.

SELECTED BIBLIOGRAPHY

Bernard, Zoë. "23andMe CEO Anne Wojcicki's Defense of Her Ex-Husband's Weird FiveFinger Shoes Is a Glimpse into How She Thinks." Business Insider, November 20, 2017. https://www.businessinsider.com/23andme-founder-anne-wojcicki-speaks-about-sergey-brin-2017-11.

Corrêa, Fernando Ribeiro. "Interview with Google's Sergey Brin." Linux Gazette, November 2000. https://linuxgazette.net/issue59/correa.html.

Malseed, Mark. "The Story of Sergey Brin. How the Moscow-Born Entrepreneur Cofounded Google and Changed the Way the World Searches." Moment, February 2007. https://momentmag.com/the-story-of-sergey-brin/.

Pangambam, S. "Google I/O 2012 Keynote—Day 1 (Full Transcript)." Singju Post, July 12, 2014. https://singjupost.com/google-io-2012-keynote-day-1-full-transcript/?singlepage=1.

Rosenberg, Scott. "Let's Get This Straight: Yes, There Is a Better Search Engine." Salon, December 22, 1998. https://www.salon.com/1998/12/21/straight_44/.

Samuelson, Kate. "Google's Co-founder Criticizes President Trump's Refugee Ban in Passionate Speech." Fortune, January 21, 2017. https://fortune.com/2017/01/31/sergey-brin-donald-trump/.

"Sergey Brin." American Academy of Achievement. Accessed February 7, 2020. https://www.achievement.org/achiever /sergey-brin/.

Sottek, T. C. "Google Co-founder Sergey Brin Joins Protest against Immigration Order at San Francisco Airport." Verge, January 28, 2017. https://www.theverge.com /2017/1/28/14428262/google-sergey-brin.

Vise, David A., and Mark Malseed. *The Google Story*. New York: Bantam-Dell, 2005.

Yurieff, Kaya. "Google Co-founders Larry Page and Sergey Brin Stepping Down as Alphabet Executives." CNN, December 4, 2019. https://www.cnn.com/2019/12/03/tech/alphabet-google -co-founder-larry-page-step-down/index.html.

FURTHER READING

BOOKS

DeAngelis, Audrey. *Google*. Minneapolis: Essential Library, 2019.
Learn how Sergey Brin and Larry Page created the world's most popular search engine and built a company that came to dominate the internet.

Di Piazza, Domenica. *Google Cybersecurity Expert Parisa Tabriz*. Minneapolis: Lerner Publications, 2018.
Tabriz holds the top spot in Google's cybersecurity department. Learn how she keeps information safe at the world's most used search engine.

Redding, Anna Crowley. *Google It: A History of Google*. New York: Feiwel and Friends, 2018.
Google is so popular that the term *google* has come to stand for any information search online. Find out about the company's birth, growth, and plans for the future.

Sichol, Lowey Bundy. *From an Idea to Google: How Innovation at Google Changed the World*. New York: Houghton Mifflin Harcourt, 2019.
With humorous illustrations and fascinating facts, this book tells Google's story.

WEBSITES

Google
https://www.google.com/
You can start a web search at this site and also learn about Google's many other products, including maps, Gmail, and Google Earth.

How Internet Search Engines Work
https://computer.howstuffworks.com/internet/basics/search-engine.htm
How does a search engine sift through millions of web pages to provide the information you're looking for online? This site explains the technology.

Safesearch: Google for Kids
https://www.safesearchkids.com/google-kids/
This site offers tips to help kids stay safe online.

INDEX